JHSC

D1448028

Williams,
Projects with wheels

DO NOT REMOVE
CARDS FROM POCKET

ALLEN COUNTY PUBLIC LIBRARY
FORT WAYNE, INDIANA 46802

8/94

You may return this book to any agency, branch, or bookmobile of the Allen County Public Library.

DEMCO

PROJECTS WITH

By
John Williams

Illustrated by
Malcolm S. Walker

Gareth Stevens Children's Books
MILWAUKEE

For a free color catalog describing Gareth Stevens' list of high-quality books, call 1-800-341-3569 (USA) or 1-800-461-9120 (Canada).

Allen County Public Library
900 Webster Street
PO Box 2270
Fort Wayne, IN 46801-2270

Titles in the Simple Science Projects series:

Simple Science Projects with Air
Simple Science Projects with Color and Light
Simple Science Projects with Electricity
Simple Science Projects with Flight
Simple Science Projects with Machines
Simple Science Projects with Time
Simple Science Projects with Water
Simple Science Projects with Wheels

Library of Congress Cataloging-in-Publication Data

Williams, John.
Projects with wheels / John Williams : illustrated by Malcolm S. Walker.
p. cm. -- (Simple science projects)
Rev. ed. of: Wheels. 1990.
Includes bibliographical references and index.
Summary: Explores the wheel through projects involving rollers, carts, wagons, and other devices using wheels to function.
ISBN 0-8368-0772-3
1. Wheels--Experiments--Juvenile literature. [1. Wheels--Experiments. 2. Experiments.] I. Walker, Malcolm S., ill. II. Williams, John. Wheels. III. Title. IV. Series: Williams, John. Simple science projects.
TJ181.5.W56 1991
621.8'2--dc20 91-50550

North American edition first published in MDCCCCLXXXXII by

Gareth Stevens Publishing
1555 North RiverCenter Drive, Suite 201
Milwaukee, Wisconsin 53212, USA

U.S. edition © MDCCCCLXXXXII by Gareth Stevens, Inc. First published as *Starting Technology — Wheels* in the United Kingdom, © MDCCCCLXXXX by Wayland (Publishers) Limited.
Additional end matter © MDCCCCLXXXXII by Gareth Stevens, Inc.

Editor (U.K.): Anna Girling
Editors (U.S.): Eileen Foran
Editorial assistant (U.S.): John D. Rateliff
Designer: Kudos Design Services
Cover design: Sharone Burris

Printed in the United States of America

2 3 4 5 6 7 8 9 97 96 95 94

CONTENTS

Words printed in **boldface** type appear in the glossary on pages 30-31.

ROLLERS

Long ago, people did not know about wheels. If they needed to move very big loads, they pulled them along the ground. To make it easier, they put several round poles under the load. Sometimes, the poles needed to be as large as tree trunks. The **ancient Egyptians** used rollers like this to move the giant stones that they needed for building the **pyramids**.

This boat is being pulled up the beach on rollers.

Using rollers

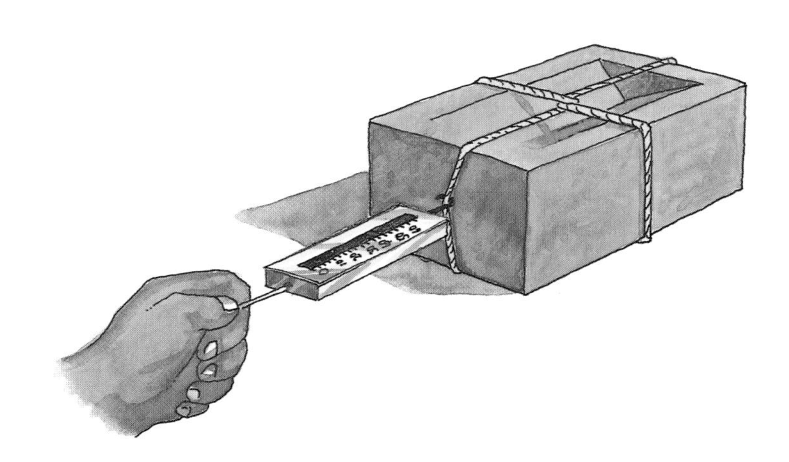

You will need:

Pencils
String
A brick
A spring balance

1. Use string to attach the **spring balance** to the brick.

2. Pull the brick along the floor by holding the ring on the end of the spring balance.

3. Look at the marker on the **scale** of the spring balance and write down the number.

4. Now put the pencils under the brick and pull it on the floor again. Is the marker at the same place on the balance as it was before? Is it higher or lower on the scale? Is the brick easier to move?

WHEEL IDEAS

Some people say that the wheel is the greatest **invention** ever. Wheels must have made life easier for everyone. Imagine a world without wheels. How could we carry all the big, heavy loads from one town to another? There would be no cars, no trucks, no trains, no bicycles — and no skateboards!

These roller skates have wide, round wheels on them.

Different kinds of wheels

Make a collection of different things that you think will make good wheels. Here are some ideas to start with:

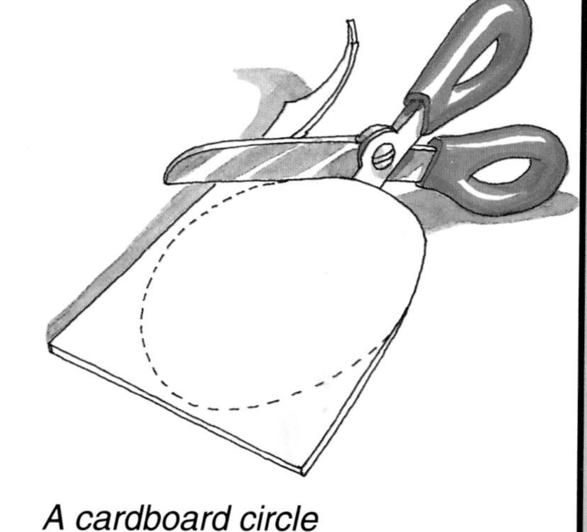

Cut out pictures of wheels. Put them in a scrapbook. See how the wheels are different from each other. Write in your scrapbook the things you learn about different kinds of wheels.

SIMPLE CARTS

You will need:

A shallow cardboard box
Cardboard
Scissors
Four paper fasteners

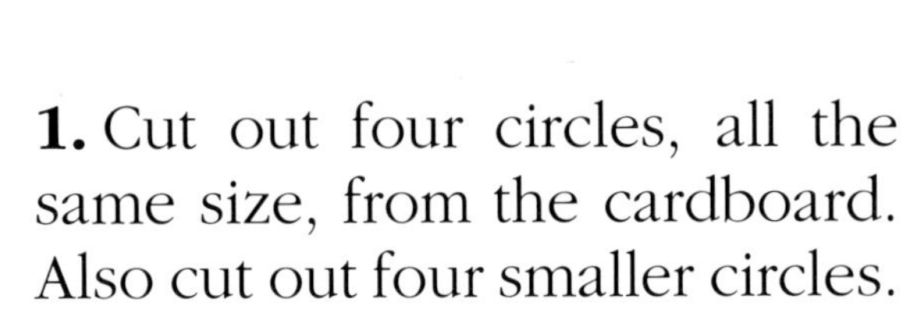

1. Cut out four circles, all the same size, from the cardboard. Also cut out four smaller circles.

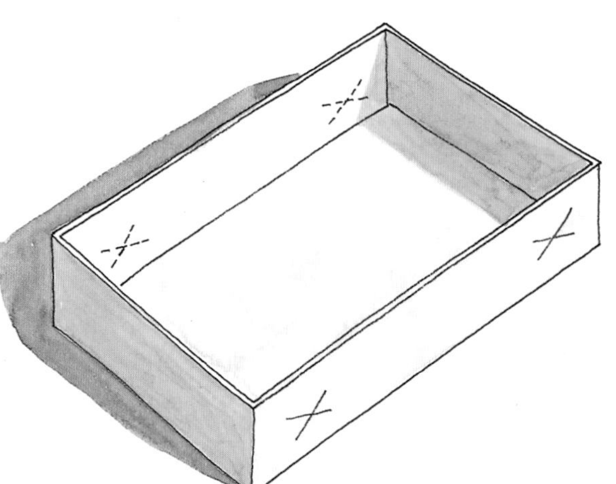

2. On one side of the box, mark the place where you will attach two of the wheels. Do the same on the opposite side of the box. Make sure your wheels line up, or your cart will wobble.

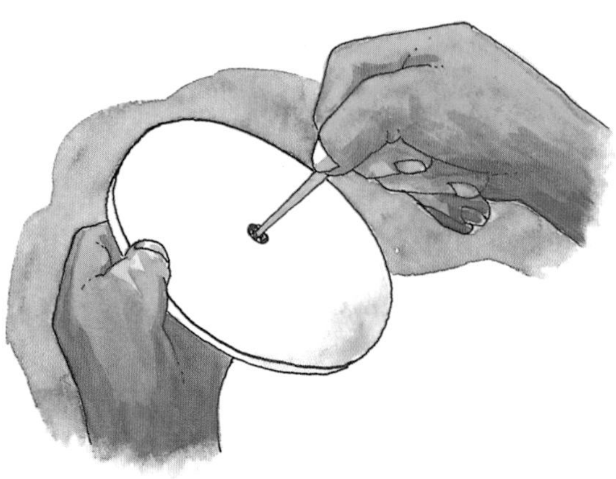

3. Make a hole in the middle of the large cardboard wheels, big enough for them to spin around on the paper fasteners.

4. Push a paper fastener through a small cardboard circle, then through one of your cardboard wheels, and, finally, through the side of the box, as shown. Open up the fastener so that it does not fall out.

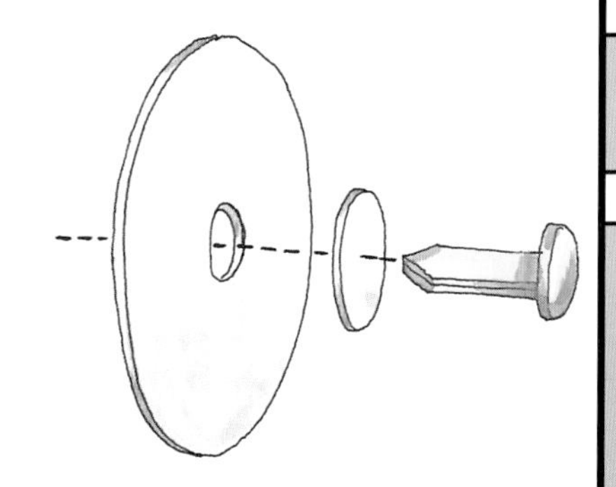

5. Attach all four wheels in this way.

This go-cart has four wheels, one at each corner, like the model cart you can make.

AXLES

The wheels that you made for the cart on page 8 were not joined together. Most wheels used on carts, trains, cars, and trucks have axles. An axle is a kind of rod that joins two wheels together so that they turn around at the same time.

Not all wheels need long axles. These racing cars do have axles, but they are very short and are attached separately.

3 1833 02473 6917

Making pairs of wheels with axles

Type One

You will need:

Two cardboard circles
A wooden stick
Glue

1. Push the stick through the center of the cardboard wheels.

2. Glue the wheels in place.

Type Two

You will need:

Two spools
A short stick
Plastic tubing
Scissors

1. Put a spool on each end of the stick, with a piece of the tubing on either side to keep them in place. Make sure the spools can spin around easily.

2. If you are not able to find any tubing, wind rubber bands around the axle instead.

WAGONS

Making a simple wagon with axles

You will need:

A small cardboard box
Two wooden sticks
Four cardboard wheels
Glue
A plank of wood

1. Make two holes on each long side of the cardboard box, as shown.

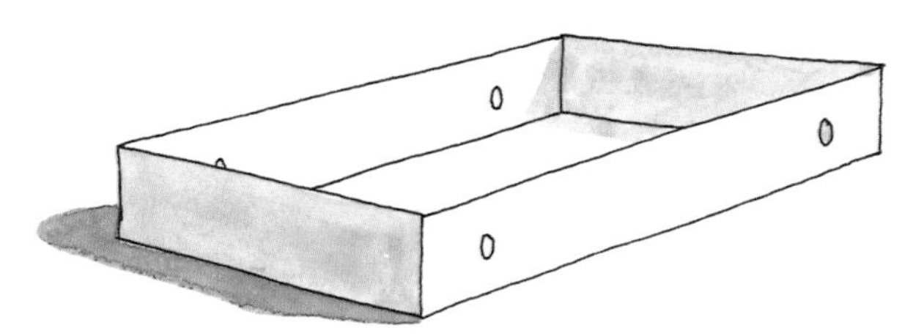

2. Push the wooden sticks through the holes. The sticks should be able to move freely in the holes.

3. Attach a cardboard wheel to each end of the sticks. The sticks should go through the center of the wheels. Glue the wheels to the sticks. The axles should be attached to the box at **right angles**.

4. Prop up one end of the plank to make a **slope**. Test your wagon by letting it run down the slope.

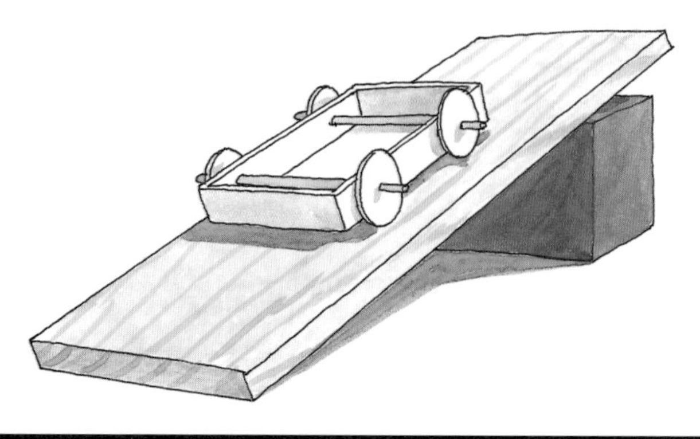

Further work

Make a wagon with its axles and wheels attached at different angles, as shown. Test your wagon on the slope. Does it run down the slope like the first wagon?

Attach different sets of wheels onto your wagon. Cut pieces of cardboard into different shapes. Here are some ideas:

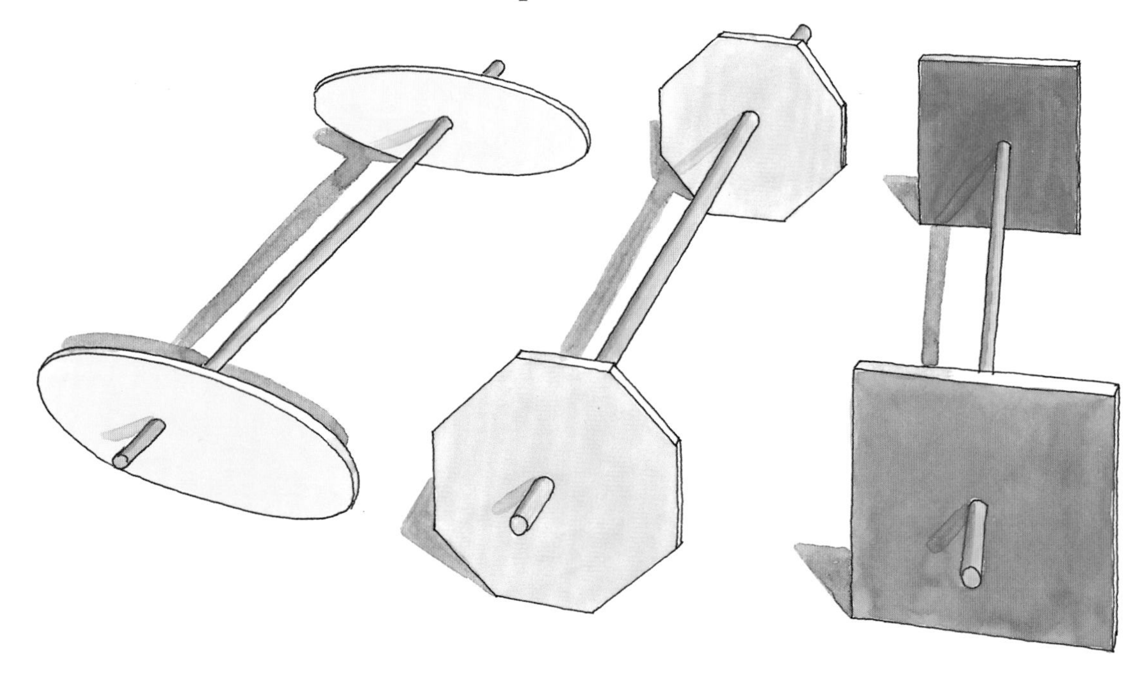

Try these different wheels on your slope. Do they work as well as round wheels?

MORE WAGONS

Making a better wagon

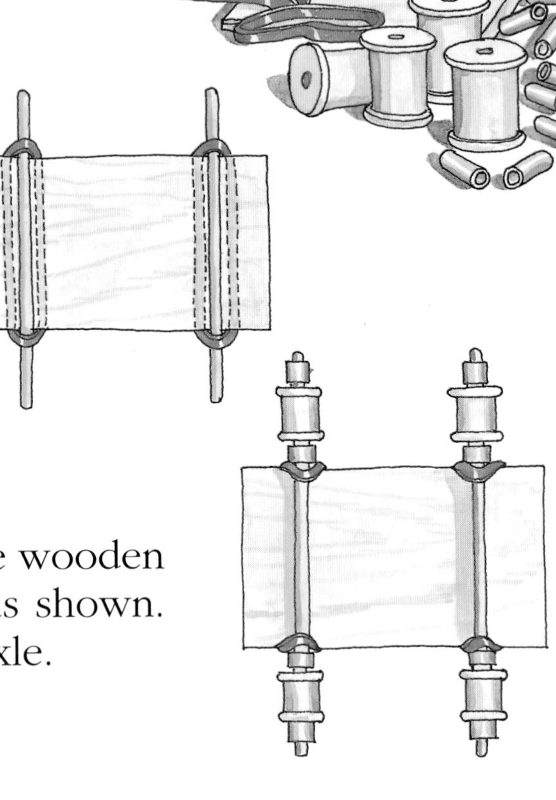

You will need:

A piece of balsa wood about 8 inches (20 cm) long and 4 inches (10 cm) wide
Two wooden sticks
Two rubber bands
Four spools
Plastic tubing
Scissors

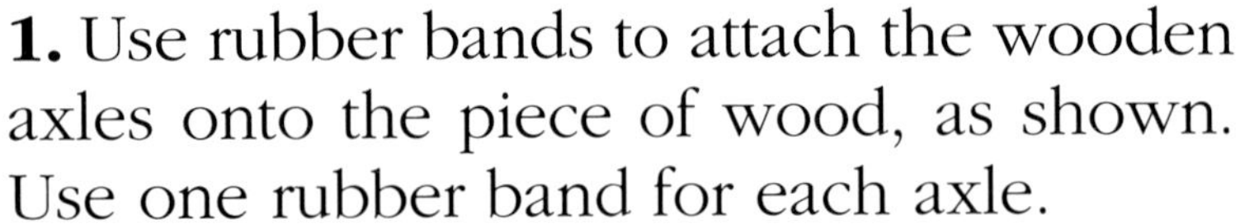

1. Use rubber bands to attach the wooden axles onto the piece of wood, as shown. Use one rubber band for each axle.

2. Slide pieces of the tubing onto each end of the axles.

3. Now slide the four spools onto the axles.

4. Attach the other four pieces of the tubing onto the ends of the axles.

Further work

Test your wagon by letting it run down a slope made from a plank of wood. Always let the wagon run down the slope from the same place. How far does the wagon go?

Change the slope of your plank. Does the wagon go the same distance if the slope is different? Put some modeling clay on the wagon. How far does the wagon travel now?

Skateboards run down slopes, but they can also travel a long way on flat ground. This is because their wheels turn smoothly.

INSECT WAGONS

Insects, such as ants, butterflies, and beetles, have six legs. You can turn the wagon you made on page 14 into an insect that has six wheels instead of six legs.

This beetle is a kind of insect. Can you see its six legs?

Making a toy insect

You will need:

Your wagon (see page 14)
Two spools
A wooden stick
A rubber band
Plastic tubing
Cardboard
Pipe Cleaners
Glue
Paints

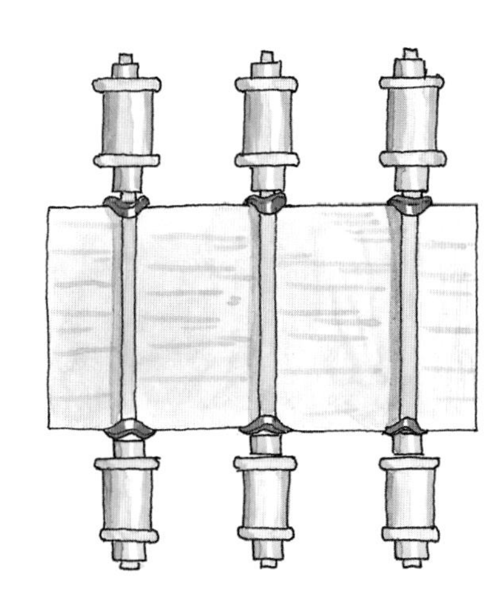

1. Use the spools, tubing, stick, and rubber band to attach another set of wheels to your wagon.

2. Cut out the shape of an insect from cardboard. Glue it to your wagon.

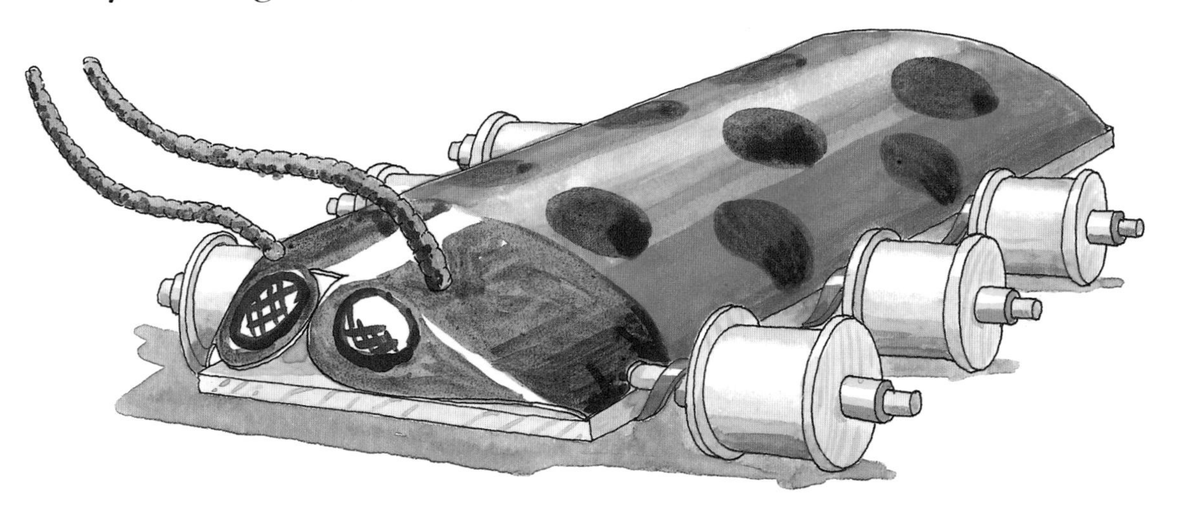

3. Paint your insect and make the **antennae** out of the pipe cleaners.

PASSENGERS

People use wheels to help them get around. Perhaps you use wheels yourself. Maybe you go to school every day in a bus or a car. Cars, buses, and trains have seats for the **passengers** to sit down. All vehicles should have **safety belts**.

This small child is wearing a special seat belt for babies.

Making a seat for a passenger

You will need:

Your wagon (see page 14)
Modeling clay and pipe cleaners
A plank of wood and a brick
Pencils and paper
Graph paper and a ruler
Rubber bands
Cardboard
Glue

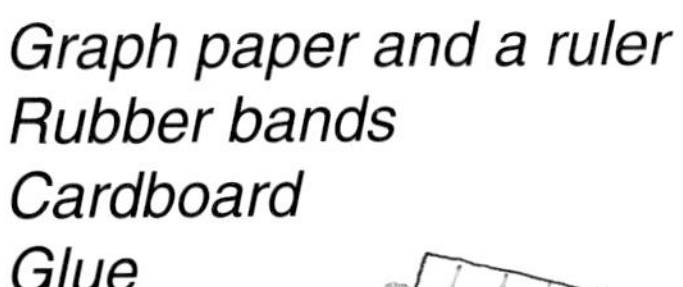

1. Construct a model passenger from some modeling clay and pipe cleaners.

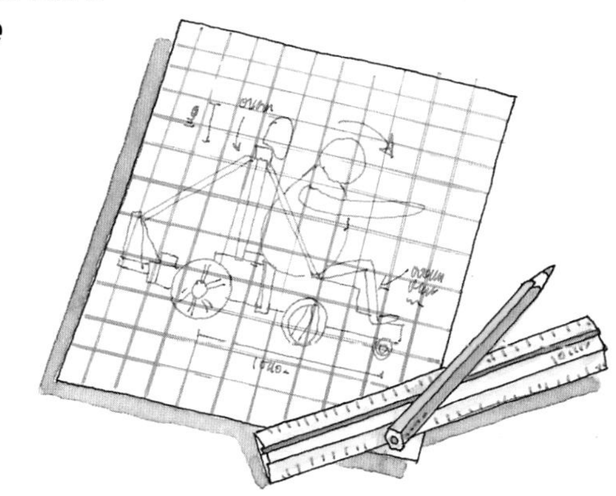

2. Think about how to make a seat for your model. How will it be attached to the wagon? How will the passenger stay in the seat?

3. On graph paper, draw a plan of the seat you are going to make. Use this as a **pattern** and make the seat out of cardboard. Attach the seat to the wagon, using glue or rubber bands.

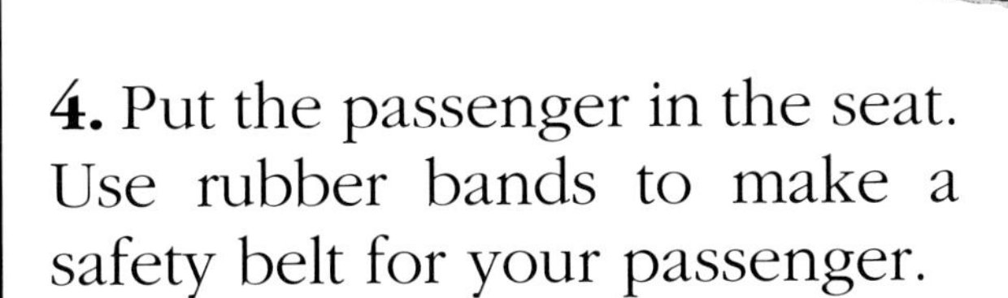

4. Put the passenger in the seat. Use rubber bands to make a safety belt for your passenger.

5. Put a brick at the bottom of the slope. What happens to the passenger when the wagon hits the brick? Now try this without the safety belt. What happens?

Making a rubber-band buggy

You will need:

A plastic dishwashing liquid bottle
A small plastic bead
Some stiff wire
A strong rubber band
A short wooden stick
A cork

1. Make a small hole in the bottom of the plastic bottle and push the rubber band through it. Use the stick to stop the rubber band from slipping through.

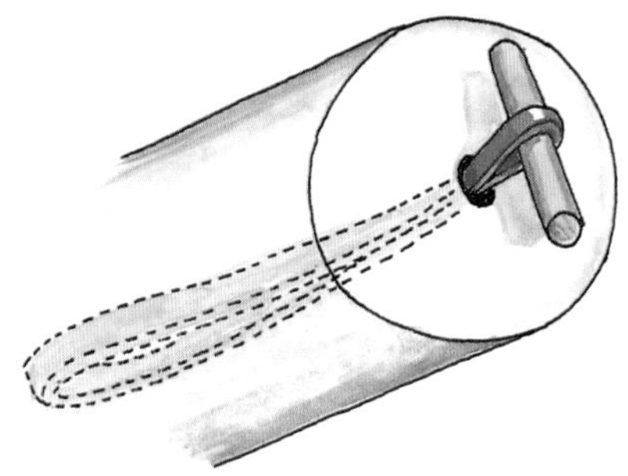

2. Remove the nozzle from the bottle. Put the wire through the bead and then through the nozzle. Bend the end of the wire into a hook shape, as shown. Attach the rubber band to the hook and then replace the nozzle.

3. Bend the wire at right angles in two places. For safety, put a cork on the end of the wire.

4. Wind up the rubber band by turning the wire around the bottle. Put your buggy on the floor and watch it move.

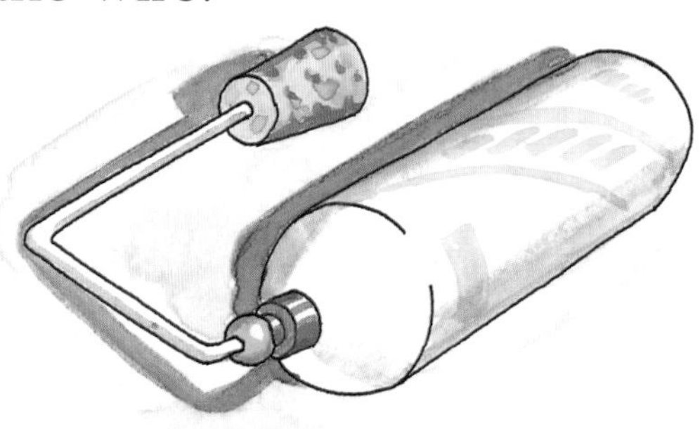

Further work

You can make a mini merry-go-round from your buggy. Make a horse from colored cardboard. Attach it to the free end of the wire. Hold the plastic bottle upright. Wind up the rubber band by turning the wire, as shown. Let it go and watch the horse spin around.

You cannot see any wheels on this merry-go-round. Do you think it has any wheels underneath to help it spin around?

COGWHEELS

Some wheels turn around but do not go forward. **Cogwheels** are special wheels that help machines work. Cogwheels are found in watches and old clocks. Some cogwheels are connected by special chains. This kind of cogwheel can be found on a bicycle, between the pedals and the back wheel.

All bicycles have cogwheels. The cogwheels are connected to each other by a chain. If you have a bicycle, look to see which cogwheel is bigger. Is it the one near the pedals or the one on the back wheel?

Making cogwheels

You will need:

Cardboard
Paper fasteners
Glue
A junior hacksaw
Popsicle sticks
Scissors

WARNING:
Always ask an adult to help you use a hacksaw.

1. Use the hacksaw to cut some Popsicle sticks in half. You will need about ten pieces.

2. Cut out a circle from cardboard.

3. Lay the pieces of wood on the circle of cardboard. Their corners should just touch each other at the center. Glue them to the cardboard.

4. Now make a smaller cogwheel in the same way. You will need about six pieces.

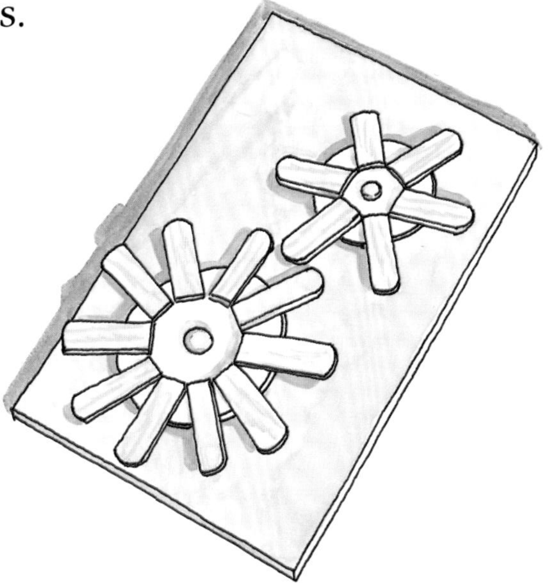

5. Push paper fasteners through the centers of the cogwheels and attach them to a large piece of thick cardboard. The cogwheels should just touch each other.

6. Move one cogwheel around. See how it makes the other one move. Move the big wheel around one whole turn. Does the small cogwheel move around more than one turn?

WATER WHEELS

Water wheels have been used for thousands of years. They are usually large wheels with paddles, or flaps, around them. Running water in a river or stream moves the paddles and turns the wheel. This movement creates energy that is then used to make machinery work.

This old water wheel was used to pump out water from underground mines, to stop them from flooding.

Making a simple water wheel

You will need:

Two Popsicle sticks
A short, round stick
A rubber band
A spool
A junior hacksaw
Plastic tubing
Scissors

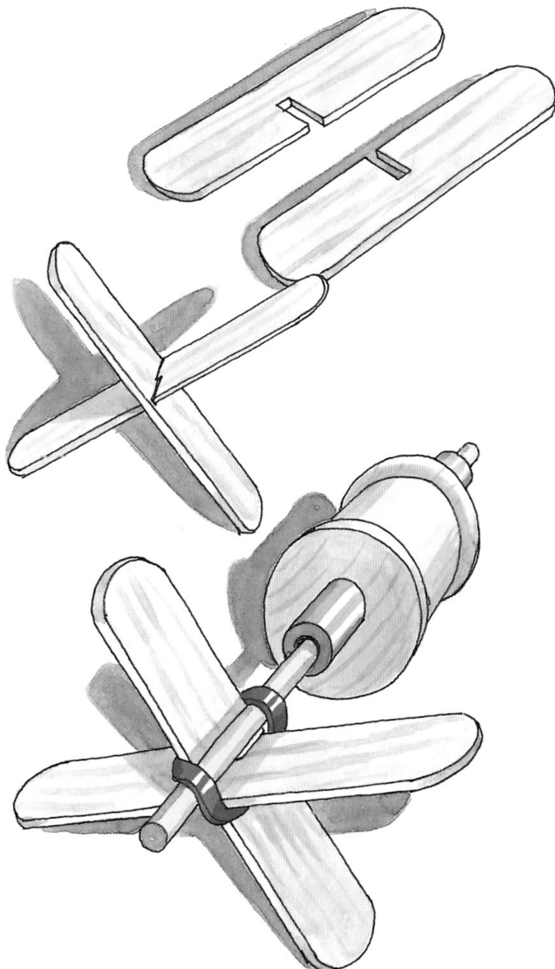

1. Cut a small notch halfway on each Popsicle stick, as shown. Slot the Popsicle sticks into each other, creating an X-shape.

2. Use a rubber band to attach the Popsicle sticks onto the end of the round stick. Slide the spool on to the other end, using the tubing on each side to hold the spool in place.

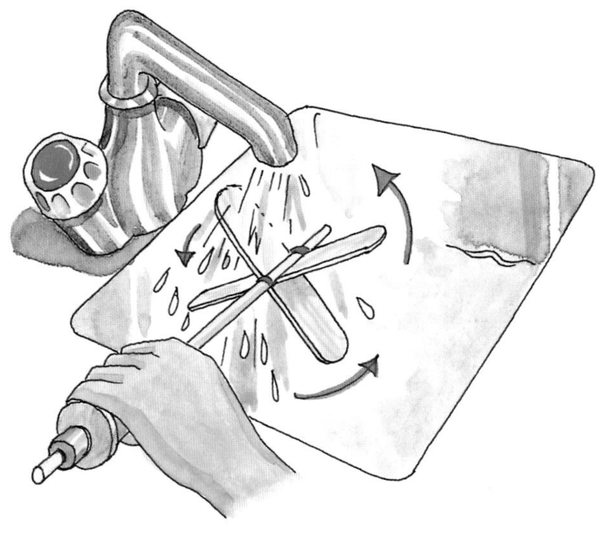

3. Hold your finished water wheel under a faucet in the sink. You should hold on to the spool so that the wheel can spin freely.

4. Slowly turn on the faucet. How fast does the water need to run before the water wheel begins to spin?

Making a water wheel that will lift a load

You will need:

Two plastic dishwashing liquid bottles
Rubber bands
String
Two round sticks
A spool
Modeling clay
A stand and clamp
Scissors

1. Make a hole at the end of one of the bottles and push one of the sticks through until it comes out of the neck. Make sure the bottle can spin freely.

2. Cut out six T-shaped pieces from the other bottle, as shown. Fold up the "stem" of each T-shape so that it sticks out.

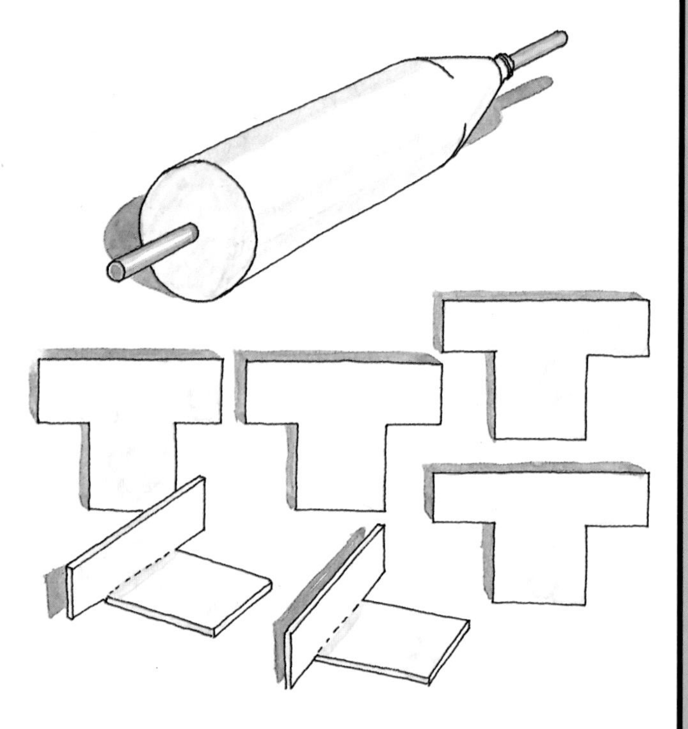

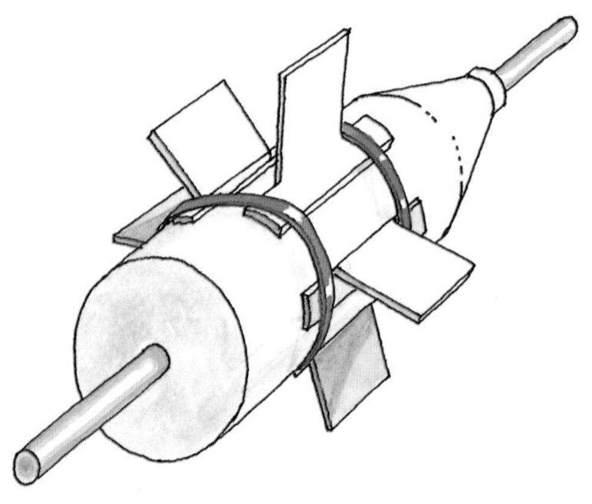

3. Use rubber bands to attach the six pieces around the bottle.

4. Hold on to each end of the stick and put the water wheel under the faucet. How fast does the water need to run to make the wheel spin?

5. Make a **pulley** by putting the second stick through a spool. Use a stand and **clamp** to set up the pulley next to the sink.

6. Tie a piece of string to your water wheel and pass it over the spool pulley.

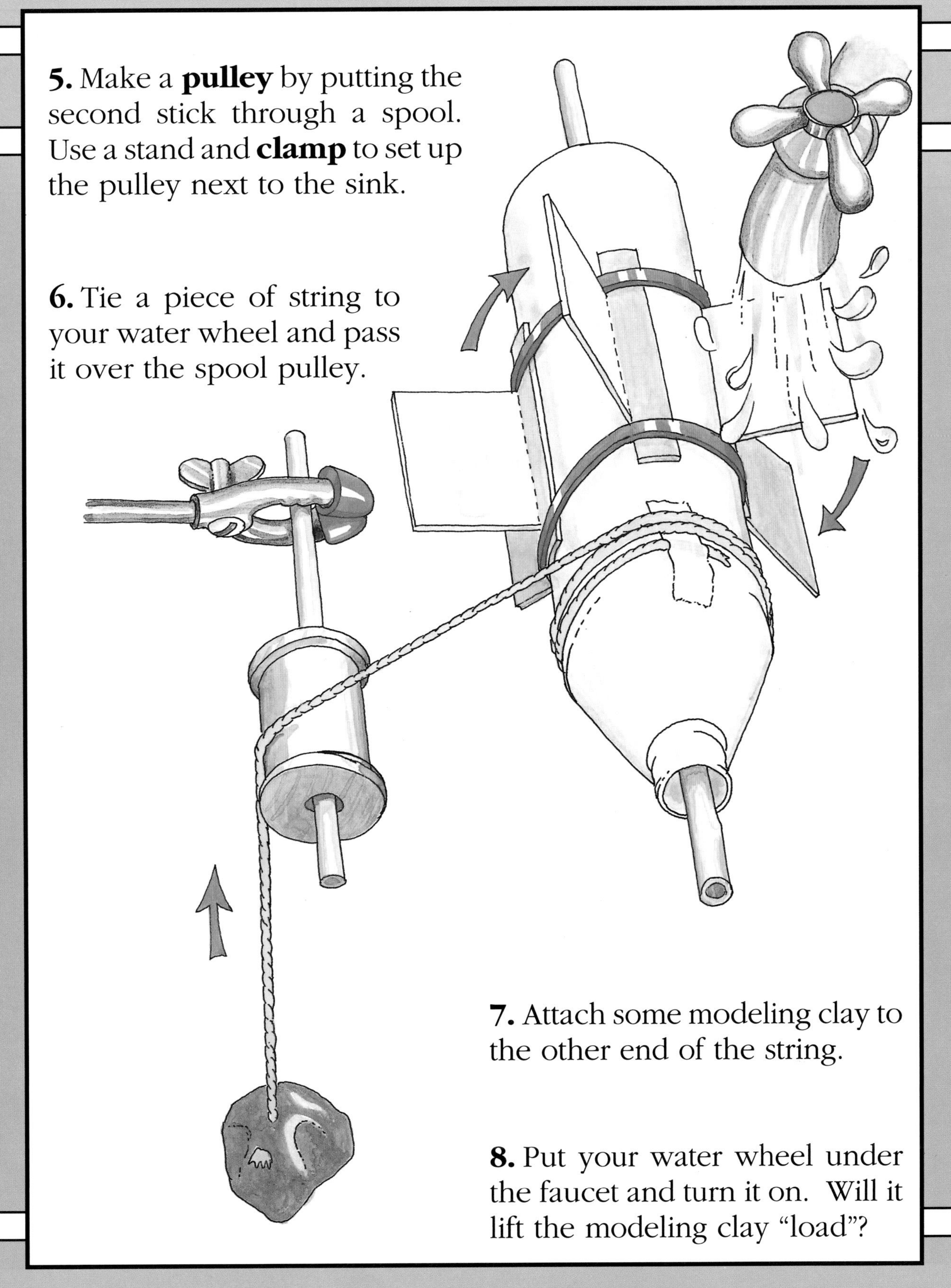

7. Attach some modeling clay to the other end of the string.

8. Put your water wheel under the faucet and turn it on. Will it lift the modeling clay "load"?

What You'll Need

More Books About Wheels

The Car: Watch It Work by Operating the Moving Diagrams! Ray Marshall and John Bradley (Viking Kestrel)
How Things Are Made. Peter Seymour (Lodestar Books)
Know Your Wheels. Henry T. Taylor (Taylor)
See How It Works: Cars. Tony Putter (Macmillan)
Weird Wheels. Alain Chirinian (Julian Messner)
Wheels. Julie Fitzpatrick (Silver Burdett Press)
Wheels: A Pictorial History. Edwin Tunis (Crowell Junior Books)

More Books With Projects

Amazing Experiments. A. Vowles (Cyril Hayes Press)
Creative, Hands-on Science Experiences. Jerry DeBruin (Good Apple)
Fifty Science Experiments I Can Do. Nancy Sundquist and Susannah Brin (Price Stern Sloan)
Make It with Odds and Ends! Felicia Law (Gareth Stevens)
The Science Book. Sara Stein (Workman)
Science Fun with Toy Cars and Trucks. Rose Wyler (Julian Messner)
Scienceworks: Sixty-Five Experiments That Introduce the Fun and Wonder of Science. Ontario Science Centre Staff (Addison-Wesley)
Smithsonian Science Activity Book. Megan Stine and others (Galison)
Two Hundred and Twenty Easy-to-Do Science Experiments forYoung People. Muriel Mandell (Dover)

Places to Write for Science Supply Catalogs

Adventures in Science
Educational Insights
19560 Rancho Way
Dominguez Hills, California 90220

The Nature of Things
275 West Wisconsin Avenue
Milwaukee, Wisconsin 53203

Delta Education
P. O. Box M
Nashua, New Hampshire 03061

Edmund Scientific
101 East Gloucester Pike
Barrington, New Jersey 08007

Nasco Science
901 Janesville Road
Fort Atkinson, Wisconsin 53538

Macmillian Science Company
8200 South Hoye Avenue
Chicago, Illinois 60620

GLOSSARY

ancient Egyptians
The people who lived in Egypt 2,000 to 5,000 years ago. They had one of the earliest civilizations known.

antennae
The pair of feelers on the head of an insect.

clamp
A device for holding objects firmly in place.

cogwheels
Interlocking wheels with "teeth," or pegs, on the outside.

insects
Small six-legged animals, such as ants, flies, and wasps.

invention
The discovery of a new way to do things, or the creation of a new type of machine.

passengers
Anyone, except the driver, who rides in a vehicle.

pattern
A plan or design that can be followed to make something.

pulley
A wheel with a rope around it. Pulleys make heavy objects easier to lift.

pyramid
A solid object with a flat base and three or more sides shaped like triangles that meet in a point at the top.

right angle
An angle is a corner where two lines meet. A right angle is a square corner, like the corner of a book or box.

safety belt
Straps that are designed to hold drivers and passengers safely in their seats.

scale
A line of regular marks used for measuring.

slope
An upward or downward slant, like a wheelchair ramp.

spring balance
A machine that measures how strong a pull is or how much something weighs. The hanging scales for produce in most supermarkets are spring balances.

Picture acknowledgements
The publishers would like to thank the following for allowing their photographs to be reproduced in this book: Cephas Picture Library, p. 22; Eye Ubiquitous, p. 21; Chris Fairclough Colour Library, pp. 6, 10, 18; Oxford Scientific Films, p. 16; PHOTRI, p. 15; Topham, pp. 4, 9; Zefa, p. 24. Cover photography by Zul Mukhida.

INDEX